Christoph Bruns

Trade Fairs as Temporary Clusters in Europe?

GRIN Verlag

Bibliografische Information der Deutschen Nationalbibliothek:

Die Deutsche Bibliothek verzeichnet diese Publikation in der Deutschen National-
bibliografie; detaillierte bibliografische Daten sind im Internet über http://dnb.d-
nb.de/ abrufbar.

Imprint:

Copyright © 2011 GRIN Verlag GmbH
Druck und Bindung: Books on Demand GmbH, Norderstedt Germany
ISBN: 978-3-656-32929-9

This book at GRIN:

http://www.grin.com/en/e-book/205776/trade-fairs-as-temporary-clusters-in-europe

UNIVERSITÄT ZU KÖLN

Wirtschafts- und Sozialgeographisches Institut

The Economic Geography of the European Union

Sommersemester 2011

Trade Fairs as Temporary Clusters in Europe?

Christoph Bruns

OUTLINE

TABLE OF FIGURES   III

TABLE OF CONTENTS

TABLE OF FIGURES

1. Introduction

In our globalized world trade fairs are important events for firms to communicate and/or sell their products and services to a global audience. But besides this original aim of trade fairs, scientists go further and focus on the effects, which these temporary and spatial accumulations of professionals from the same or equal industries might have. According to that, a particular focus lies on inter-organizational learning processes, like they can be found in clusters. Thus, the central question of this report derives, whether in the European context trade fairs can be seen as temporary clusters, which would imply that trade fairs became central nodes connecting global economy.

This issue is important, since both participating in trade fairs and searching for adequate interaction partners are costly and time intensive processes. If trade fairs provided equal benefits as temporary clusters, organizing and participating entities would expend more effort on planning and conducting the time before, during and after the trade fair. Apparently, benefiting from new knowledge pools is at least an important aspect in times of increasing innovation velocity.[1]

Finding a clear answer for the problem is not trivial, since the majority of available literature focuses on trade fairs in the context of a communication instrument in firm's marketing mix. Furthermore there are difficulties to maintain a particular focus on Europe, since globalized world economy meets on international flagship trade fairs. Another aspect is the difficulty of measuring qualitative and quantifiable effects of spatial proximity in clusters, which additionally could be implemented to compare different forms of temporary clusters.

At the outset this report provides an overview about the concepts of trade fairs and temporary clusters. Further, it focuses on the comparison of essential characteristics between these organizational structures. Especially the general structure and the matter of how information and knowledge is created will be observed. Subsequently, brief implications for the European Economy are revealed. Finally, in the conclusion, this report gives an adequate answer to the initial question to what extent trade fairs can be seen as temporary clusters in Europe.

---

[1] vgl. Grape, 2011, S. 1

2. Trade Fairs

According to the Association of the German Trade Fair Industry, a trade fair is defined as a temporally limited, recurrent market event on which a plurality of companies exhibits the essential offer of at least one economic sector and distributes predominantly by sample to commercial consumers.[2] Trade fairs create a special environment where geographical proximity and face-to-face contact enable members of an industry to exchange information and to learn about new developments concerning markets, products and innovations.[3][4]

Trade fairs can be differentiated with respect to two attributes (Figure 1). These are on the one hand the trade fair's geographic coverage and on the other hand the trade fair's market coverage.

Geographic Coverage	Market Coverage	
	Vertical	Horizontal
Regional	Regional Vertical Shows	Regional Horizontal Shows
National	National Vertical Shows	National Horizontal Shows

Figure 1: Types of Trade Fairs (Source: Bello/Barczak, 1990, S. 51)

The geographic coverage refers to the attendees' travel distance to the trade fair. In terms of a regional trade fair, almost half of the attendees travel less than 50 miles, while the majority of a national trade fair overcomes a distance of more than 400 miles.

---

[2] vgl. AUMA, 2010, S. 25
[3] vgl. Bello/Barczak, 1990, S. 48
[4] vgl. Bathelt/Schuldt, 2008, S.853

Since the typology bases on the huge geographical dimensions of the USA and regarding the European context of this report, the characteristics behind the concept of "national trade fairs" are also valid for "international trade fairs".

Further, the unequal travelling requirements reason a different mix of professional groups between the two types of trade fairs. Regional trade fairs draw a variety of low-level operating personnel, mostly working in technical jobs. This relation is reversed for national trade fairs, where a higher percentage of senior managers working in professional job functions can be observed. Regarding the market coverage, trade fairs considered as vertical are characterized by a single industry with one specialized focus. Besides top managers, this kind of trade fair also draws low-level managers and blue-collar workers. Horizontal trade fairs have the distinction of being focussed on a multitude of unrelated industries. The attendees of those trade fairs reflect a high percentage of top-level managers, who typically have procurement functions as e.g. selecting new suppliers.[5]

3. Temporary Clusters

The concept of a cluster was introduced by Porter and defines it as a geographic and sectoral concentration of interconnected companies and institutions. Further clusters often extend downstream to channels, customers and sideways to complementary producers and to companies in, by technologies or common inputs, related industries.[6] Additionally, governmental and other institutions can be implied in clusters and thus provide specialized support for different fields like education, information, research or technique. Thus clusters are furthermore characterized by spatial proximity generating local knowledge spill-overs, leading to competitive advantages for the co-located firms.[78] The difference between a temporary cluster and a permanent cluster is, besides the limited time horizon of a temporary cluster, its intensified form. Hence, Maskell et al. describe it as a *"short-lived hotspots of intense knowledge exchange, network building and idea generation"*.[9]

---

[5] vgl. Bello/Barczak, 1990, S. 50
[6] vgl. Bresnahan et al., 2001, S. 836
[7] vgl. Porter, 1998, S. 78
[8] vgl. Bresnahan et al., 2001, S. 836
[9] vgl. Maskell et al., 2006, S. 997

3.1 Structure

Like trade fairs, clusters can be split into a horizontal and a vertical dimension. The horizontal one consists of those firms that produce similar goods and compete with each another. Although a close contact between the firms is not necessary, the proximity to competitors provides the opportunity to compare and monitor the rival's products and its qualities. In an industrial cluster, where every firm's production also underlies the same local conditions, companies can also estimate the amount of production costs and thus increase the quality of information generated by effectively comparing the own performance with the competitors' performances. Hence, this procedure evokes high incentives for co-located firms to continuously innovate.[10] The vertical dimension of a cluster is, in terms of growth, positively correlated to the existing variety of the horizontal dimension.[11] That implies, that an established accumulation of similar or complementary producing firms reasons the upcoming demand for specialized supply and services. Further, providers of these specific demands also move, at least with one small business entity, to the clustered region and thus establish its vertical axis.[12]

3.2 Local Buzz

A resulting phenomenon from that spatial proximity in clusters is the so-called "buzz", which already was hinted by Marshall as a non-moveable "special atmosphere" within clusters.[13] Thus, it is the co-location of people and firms within the same industry, place or region and further the resulting face-to-face contact, which create an "information and communication ecology". The buzz relies on that issue and hence consists of a specific flow of information and its continuous update. Taking the knowledge typology elaborated by Polanyi into account, previous literature stated that clusters do provide opportunities for the transfer of the sticky, tacit forms of knowledge. This seems quite clear, since the appropriation of the non-codeable tacit knowledge is just feasible through making personal experiences and hence heavily depends on the social surrounding.[141516] That means that organized or accidental meetings, same interpretation

---

[10] vgl. Bathelt et al., 2004, S. 36
[11] vgl. Marshall, 1920, S. 225
[12] vgl. Bathelt et al., 2004, S. 37
[13] vgl. Marshall, 1927, S. 284
[14] vgl. Gertler, 2003, S.78
[15] vgl. Polanyi, 1966, S.4ff.
[16] vgl. Bathelt et al., 2004, S. 32

schemes of new knowledge inputs and shared cultural traditions within a particular technology field, excite inter-firm learning processes. This is crucially important for a whole industry.[17] Finally by implementing monitoring, comparing and screening the important information of the buzz, companies can deduce further assumptions about the rivals' strategies. Obviously, this facilitates companies' future decision making and stimulates the implementation of reflexive practices within the firm.[18]

3.3 Pipelines

Although literature did not find significant empirical results about the effectiveness of localized learning, the success of those firms situated in a cluster gives causes to support the stated arguments.[19] But nevertheless in the long run the successful development of a cluster, meaning that positive growth is generated within the co-location, is dependent on external sources of knowledge. These "pipelines" are strategic partnerships of interregional and international reach and provide non-incremental knowledge flows.[20] Pipelines deliver valuable stimulative impacts and hence also improve inter-firm learning processes. But in contrast to the relatively unstructured and broad information of the local buzz, they consist of more structured information.[21] Another difference is the high amount of inter-firm trust, which allows the mutual support with high quality information.[22]

4. Trade Fairs as Temporary Clusters

4.1 Structure

Both of the initial paragraphs revealed that trade fairs and clusters can be characterized in a vertical and horizontal dimension. In terms of trade fairs the horizontal dimension describes unrelated industries, but concerning clusters, the horizontal dimension refers to different manufacturers, which produce similar or complementary goods. Apparently, there is a lack of conceptual clarity, but if the vertical dimension of trade fairs by Bello/Barczak is taken into account, meaning a single industry with a specialized focus,

---

[17] vgl. Bathelt et al.,2004, S. 38
[18] vgl. Bathelt et al.,2004, S. 36
[19] vgl. Oinas, 1999, S. 364f.
[20] vgl. Owen-Smith/Powell, 2002, S. 3f.
[21] vgl. Bathelt et al., 2004, S. 33
[22] vgl. Maskell/Malmberg, 1999, S. 19

5

then it becomes clear that it fits with the horizontal dimension of a cluster. Hence, the structure of a cluster is also reflected in vertical trade fairs. This can also be revealed in the mixture of professional groups and functions. On the one hand, there is the specialized focus on a single industry, which also draws suppliers to the trade fair; on the other hand the attendees are not exclusive from top-level management or top-level engineering positions, they are also lower-level employees and technical workers.

In terms of the structure horizontal trade fairs do not correlate that well with clusters, due to the presence of a multitude of unrelated industries with a less depth along the value chain. Furthermore, the high percentage of top-level managers does not reflect the constellation of the high variety of professional groups in clusters, where every company's employee is located within.

So finally vertical trade fairs and clusters share the same narrow industry focus and also the same represented constellation of professional groups with equal functions.

4.2 Information and Knowledge Creation

Both trade fairs and clusters have the quality of bringing those companies together, which normally would struggle with each other for conquering the superior position in their competitive market. But the effects of spatial proximity in these co-locations enable every participant to benefit from each other's presence.

4.2.1 Monitoring and Comparing

First, there is the opportunity for continuous monitoring of the competitors' products, product quality. Due to the same local conditions, companies can also estimate the amount of production costs and thus increase the quality of information generated by effectively comparing the own performance with that of the competitors.[23] This sharpness of insights can rather be provided by a trade fair, since it is a place without any production facilities and thus inconclusively reveals any information about cost structures, which could serve as point of reference. Nevertheless, monitoring and comparing the rival's booth at a trade fair already provides valuable information.

---

[23] vgl. Bathelt et al., 2004, S. 36

4.2.2 Buzz and Global Pipelines

For generating a proper overview, firms also need to seek relevant information within the local buzz. Since it refers to an "information and communication ecology" created by spatial proximity and face-to-face contact, the buzz is also present on trade fairs. That implies that also trade fair attendees continuously conduce to and benefit from the diffusion of information. According to that, trade fairs support both the transfer of explicit and also tacit knowledge like it is the case in temporary clusters.[24] Maskell et al. also argue that knowledge generated at trade fairs is similar to the incremental improvements that co-located rivals reveal in the cluster.[25]

Another important characterization of clusters is the dependence on external knowledge sources, implemented by global pipelines. They generate growth and consequently enable the cluster to remain competitive. The dynamic information ecology and the multitude of represented firms at trade fairs support the formation of external pipelines.[26] But with regard to the initial trade fair typology (Figure 1) it becomes clear that only international trade fairs can provide an access to global knowledge pools and markets, since they attract an agglomeration of agents from all over the world, representing the industry, science, institutions or international media. These attendees create a temporary hotspot microcosm on a global level. Since regional trade fairs do not draw attendees from far out regions, its concept does not fit to the idea of a temporary cluster.

It is quite obvious that the global input also has an influence on the buzz's nature. Through international participation at a trade fair its buzz becomes a global one, which still contains explicit and tacit knowledge. Hence, it has a higher amount of relevant information than the local buzz of regional trade fairs. That implies that the monitoring and observation process conducted by firms to filter relevant information out of the buzz might become more difficult. But its multidimensional and global structure also enables firms to detect more business opportunities e.g. by identifying new international partners. A further aspect regarding the relationship between global pipelines and the local buzz is assumed by Bathelt et al., who see a mutual reinforcing connection. The higher the amount of global pipelines going out from co-located firms, the more

---

[24] vgl. Bathelt/Schuldt, 2008, S. 864
[25] vgl. Maskell et al., 2006, S. 1007
[26] vgl. Maskell et al., 2006, S.1002

information about markets or technologies get into the cluster and hence increase the dynamic of its buzz. That implies benefits for cluster firms and further also for their strategic partners on the other side of the pipeline.[27]

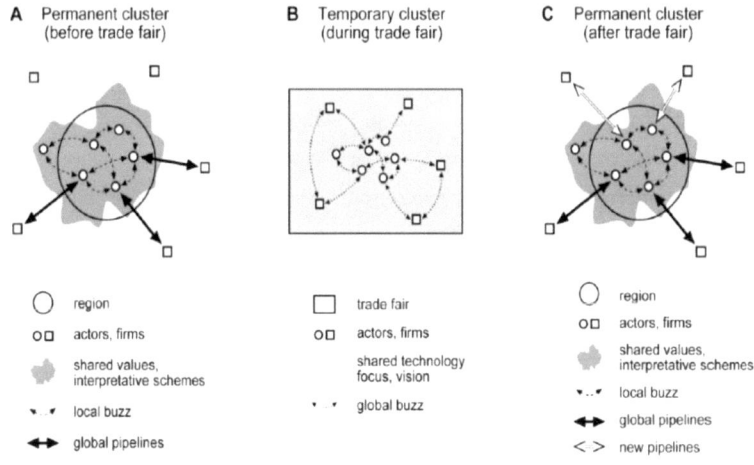

Figure 2: Pipeline Creation through International Trade Fairs
(*Source: Bathelt/Schuldt, 2008, S.856*)

Nevertheless identifying, selecting, approaching and interacting with new partners is complicated and time consuming, which makes establishing global pipelines become a costly process.[28] But if firms do participate in international trade fairs, specific commitments and additional investments to initiate contact with other like-minded people are not necessary. Hence scouting for partners and subsequently establishing of new pipelines at trade fairs is relative easy (Figure 2), since it does not involve risk and extra costs.[29] According to the specific and valuable content of global pipeline communication, trust is a vital and necessary ingredient for the relationship. Since trade fairs normally take place in different cities, meaning that attendees are generally not surrounded by the stress of daily routine, they easier provide the opportunity to agree on informal meetings in the evening to reduce uncertainties and develop trust. For establishing a long-term relationship based on trust, like in permanent clusters, for

---

[27] vgl. Bahtelt et al., 2004, S. 41
[28] vgl. Maskell et al., 2006, S. 998
[29] vgl. Bathelt/Schuldt, 2008, S. 864

intensifying or just for maintaining this network, regular attendance of trade fairs is important. Furthermore latent pipelines can then be activated, when they are needed.[30][31]

Finally the question comes up, whether global pipelines also enable firms to transfer their accumulated tacit knowledge or if these information flows are limited to explicit knowledge. According to that, literature of Economic Geography is not that clear. On the one hand some authors argument that knowledge becomes increasingly ubiquitous and thus the competitive advantage of local learning gets less meaningful.[32] On the other hand there are authors who argue that tacit knowledge remains disembodied and thus, sticky information cannot be transferred.[33] But they all agree that the benefits resulting from spatial proximity and its special atmosphere as explained in previous chapters are still the reason for competitive advantages.

Further, according to continious improvements of communication-technology and decreasing transaction costs the importance of permanent geographicallly proximity might decrease.[34] But routines and social practices remain vital for inter-firm knowledge creation. That implies that permanent clusters are not necessary, since temporary clusters are sufficient to establish these pipelines based on trust.

5. Implications for the European Economy

According to the high amount of international trade fairs, periodically taking place in Europe, for European firms it is less cost intensive and hence more likely to participate. That implies, that there must exist a positive influence on European Economy, since European firms and its surroundings have more opportunities to benefit from the stated advantages of temporary clusters. Hence, it also should be easier for them to maintain and establish more external pipelines, which can serve as important knowledge source. Going more into detail, especially for companies with facilities in Germany, where the majority of all trade fairs within Europe takes place, there should be a positive impact. Regarding these arguments and according to increasing numbers of exhibition spaces in Asia/Pacific, European economic policy should provide optimal trade fair conditions to

---

[30] vgl. Maskell et al., 2006, S. 1002
[31] vgl. Bathelt/Schuldt, 2008, S. 855f.
[32] vgl. Maskell/Malmberg, 1999, S. 9ff.
[33] vgl. Asheim, 2002, S. 117
[34] vgl. Porter, 1998, S. 90

9

preserve this positive influence and hence keep its dynamics for innovation and growth in the European Union.[35]

6. Conclusion

Previous chapters stated essential characteristcs of both trade fairs and clusters. It could be revealed that there are indeed certain overlapping aspects.

Initially it was underlined that the structure of a temporary cluster is also reflected in vertical trade fairs, since they share the same narrow industry focus and the same represented constellation of professional groups with equal functions. Consequently both trade fairs and temporary clusters create a local microcosm, in which an entire world market of an industry is compressed. Thus, various opportunities for firms to benefit from their participation are provided. Monitoring and comparing enable firms to evaluate the own performance in the competitive context. Furthermore face-to face contact and spatial proximity evoke the buzz and further knowledge spillovers, which also consist of tacit knowledge. Hence, an improvement or a more adequate alignment of products and strategies across different markets becomes feasible.

Under the restriction that trade fairs are international construed, they also provide the opportunity to find new partners, in order to establish global pipelines. These external linkages are vital for a sustainable development of clusters and can be maintained by developing a trust-based relationship. Regular attendance at trade fairs and informal meetings subsequently support trust building and decrease uncertainties.

According to this argumentation line and also taking the structural constraints for trade fairs into account, I finally come to the conclusion to answer the initial question in the affirmative. International, vertical trade fairs are central nodes that do connect global economy and provide various stated benefits for every participant. With regard to my stated results, they are temporary clusters and thus short-lived hotspots of intense knowledge exchange, network creation and idea generation.[36]

---

[35] vgl. UFI, 2010, S. 7ff.
[36] vgl. Maskell et al., 2006, S. 997

Bibliography

AUMA - Ausstellungs- und Messe-Ausschuss der Deutschen Wirtschaft e.V. (2010): Erfolgreiche Messebeteilung- Teil 1 Grundlagen. Online im Internet unter: http://www.auma.de/_pages/d/17_Publikationen/1701_Uebersicht/17010107_Erfolgreic heMessebeteiligung1.aspx [Stand: 10.05.2011].

Asheim, B.T. (2002): Temporary organisations and spatial embeddedness of learning and knowledge creation. In: Geografiska Annaler, Jg. 84, H. 2: S. 111–124.

Bathelt, H.; Malmberg, A.; Maskell, P. (2004): Clusters and knowledge: local buzz, global pipelines and the process of knowledge creation. In: Progress in Human Geography, Jg. 28, H. 1, S. 31-56.

Bathelt, H.; Schuldt, N. (2008): Between Luminaires and Meat Grinders: International Trade Fairs as Temporary Clusters. In: Regional Studies, Jg. 42, H. 6, S. 853-868.

Bello, D. C.; Barczak, G. J. (1990): Using Industrial Trade Shows to Improve New Product Development, Journal of Business and Industrial Marketing, Jg. 5, H. 2, S. 43-56.

Breshnahan, T.; Gambardella, A.; Saxenian, A. (2001): 'Old Economy' Inputs for 'New Economy' Outcomes: Cluster Formation in the New Silicon Valley. In: Industrial and Corporate Change, Jg. 10, H. 4, S. 835-860.

Gertler, M. S. (1995): 'Being there': proximity, organization and culture in the development and adoption of advanced manufacturing technologies. In: Economic Geography, Jg. 71, H. 1, S. 1-26.

Grape, C. (2011): Die Welt in der wir leben: Akademische Einwürfe zum Verständnis unserer Zeit. Online im Internet unter: http://www.uni-duesseldorf.de/home/startseite/news-detailansicht/article/oeffentliche-ringvorlesung-studium-universale.html [Stand: 18.05.2011].

Marshall, A. (1920): Principles Of Economics. 8. Auflage. Philadelphia: Porcupine Press.

Marshall, A. (1927): Industry and trade- A study of industrial technique and business organization and their influences on the conditions of various classes and nations. 3. Auflage. London: Macmillian.

Maskell, P.; Bathelt, H.; Malmberg, A. (2006): Building Global Knowledge Pipelines: The Role of Temporary Clusters. In: European Planning Studies Jg. 14, H. 8, S. 997-1013.

Maskell, P.; Malmberg, A. (1999): The Competitiveness of Firms and Regions: 'Ubiquitification' and the Importance of Localized Learning. In: European Urban and Regional Studies, Jg. 6, H. 1, S. 9-25.

Oinas, P. (1999): Activity-specificity .in organizational learning: implications for analysing the role of proximity. In: GeoJournal, Jg. 49, H. 4, S. 363-372.

Porter, M. E. (1998): Cluster and the new Economics of Competition. In: Harvard Business Review, Jg. 76, H. 6, S. 77-90.

UFI -Union des Foires Internationales (2010): Global Exhibition Industry Statistics Nov. 2010. Online im Internet unter: http://www.ufi.org/media/publicationspress/2010_exhibiton_industry_statistics.pdf [Stand: 18. 05. 2011].